666 Jellybeans! All That?

An Introduction to Algebra

Think of a number.
Double it....
    We have all played this kind of game—but how many of us realized that algebra is what makes it work?
    Malcolm E. Weiss, in this inventive and amusing introduction to algebra for young children, shows that the "unknown," or X, can be the contents of a paper bag or pocket, or just the number you are thinking of.
    With a word of caution about using large numbers (they make the arithmetic in the game harder), Mr. Weiss happily leads the reader to an equation with a pretty big answer. (See the title of the book.)
    Judith Hoffman Corwin's illustrations help to clarify the problems and make 666 JELLYBEANS! ALL THAT? a handsome book to read and share.

666 Jellybeans! All That?

AN INTRODUCTION TO ALGEBRA

by Malcolm E. Weiss

illustrated by Judith Hoffman Corwin

Thomas Y. Crowell Company

New York

YOUNG MATH BOOKS

Edited by Dr. Max Beberman, Director of the Committee on School Mathematics Projects, University of Illinois

Edited by Dorothy Bloomfield, Mathematics Specialist, Bank Street College of Education

Library of Congress Cataloging in Publication Data. Weiss, Malcolm E. 666 jellybeans! all that? SUMMARY: A simple introduction to algebra for very young children. 1. Algebra—Juv. lit. [1. Algebra] I. Corwin, Judith Hoffman. II. Title. QA155.W39   512.9'042 75-9528   ISBN 0-690-00914-3 (CQR)

1 2 3 4 5 6 7 8 9 10

666 Jellybeans! All That?

An Introduction to Algebra

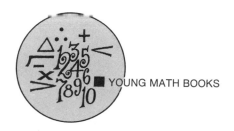

YOUNG MATH BOOKS

Think of a number from 1 to 5.
Add 1 to it.
Double your answer.
Then add 4.
Divide what you get by 2.
Now subtract 3.

What number do you end up with? Try it with each of the numbers from 1 to 5. What do you notice about your answers?

$$2$$

$$2+1=3$$
$$2\times3=6$$
$$6+4=10$$
$$10\div2=5$$
$$5-3=2$$

5

$$5 + 1 = 6$$
$$2 \times 6 = 12$$
$$12 + 4 = 16$$
$$16 \div 2 = 8$$
$$8 - 3 = 5$$

Does this number trick work with other numbers, too? Pick some other numbers and find out what happens. But watch out! If you start with a big number, you'll have to work with even bigger ones.

You might want to try 6 or 7, for example. Maybe even 10 or 20. But you probably wouldn't want to test the trick for 666.

Maybe the trick works for *all* numbers. If you could prove that, you would know that 666 would work without having to do all the arithmetic. So would any number, no matter how big.

Is there a way to prove it? Let's go back and do the trick one more time.

This time, suppose you ask a friend to think of a number. You don't know what the number is. But you want to keep track of what happens to the number at each step of the trick.

To help her keep track of the number, your friend could use a paper bag. After she thinks of a number, she takes that many jellybeans and puts them in the bag. Of course, if your friend picks a number like 666 she is going to have to find an awful lot of jellybeans. But that's her problem. Remember, you don't need to know what the number is. All you need to start with is the bag.

Now you can make a picture of what happens at each step of the trick. Adding 1 gives you the bag plus an extra jellybean. Doubling— multiplying by two—means another bag, filled with the same number of jellybeans, plus another jellybean. This makes two bags and two extra jellybeans. Then adding 4 makes two bags and six jellybeans. Taking half of that is what you would get if you shared the whole thing equally with your friend. Your share would be one bag and three jellybeans.

Taking away three jellybeans leaves you with just the bag. It doesn't matter how many jellybeans were put into the bag.

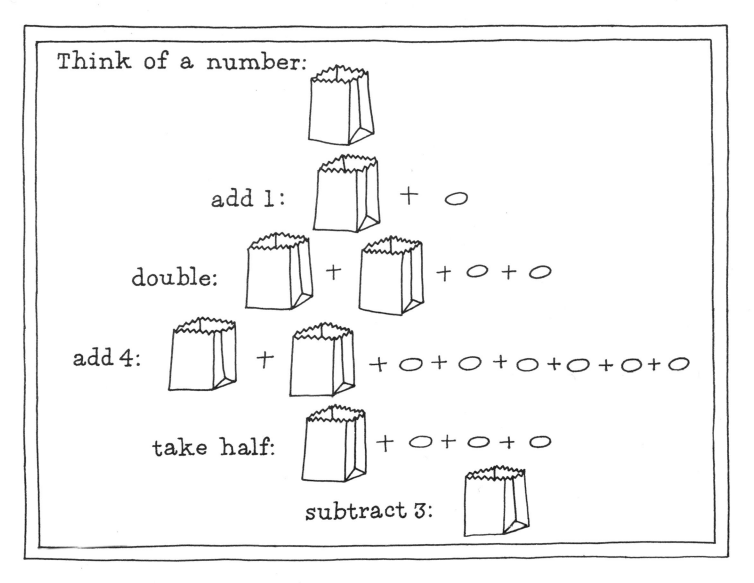

Think of a number:

add 1:

double:

add 4:

take half:

subtract 3:

If it was one, you end up with a bag holding one jellybean. If it was 666, you end up with a bag holding 666 jellybeans.

No matter what number you start with, you end up holding the bag. The bag holds—or stands for —the number your friend thought of. So you have proved that the trick works for any number. To prove it, you have used the bag in place of a number in adding, subtracting, multiplying, and dividing.

Here's another way you can use the bag like a number. Suppose you change the last step in the trick. Instead of subtracting 3, you say, "Subtract the number you first thought of."

What number does the trick end with now?
Have you turned the first trick into a new one?

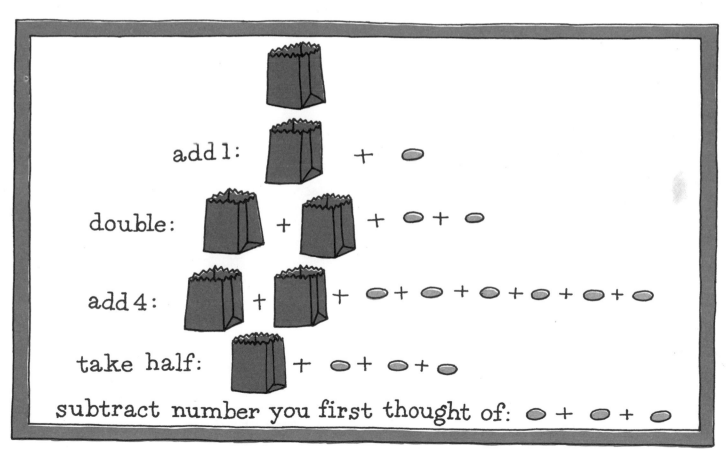

add 1:

double:

add 4:

take half:

subtract number you first thought of:

When you use the bag to stand for a number that you don't know, you are doing ALGEBRA. The bag is a SYMBOL for the unknown number.

Mathematicians don't usually draw bags to stand for unknown numbers. They use letters of the alphabet. Letters and numbers are the language of algebra. To write out our number trick in this language, you might use the letter B instead of the bag.

Then adding 1 gives $B + 1$. Can you see why doubling, or multiplying by 2, makes this $2 \times B + 2$, which mathematicians write as $2B + 2$? What would you get if you multiplied $B + 1$ by 3? By 4? By 5? By N?

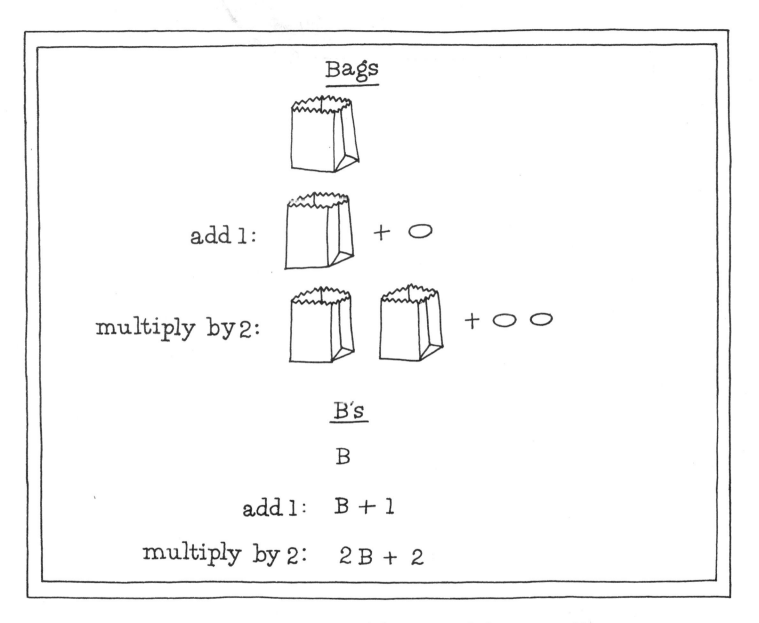

Bags

add 1:

multiply by 2:

B's

B

add 1: B + 1

multiply by 2: 2 B + 2

Can you finish the proof for the trick yourself,
using the language of algebra?

add 4 : 2B + 6

divide by 2 : B + 3

subtract 3 : B

Algebra helps mathematicians work out puzzles and problems that can't be easily solved by arithmetic. That's because in arithmetic you can only use numbers that you know about. In algebra, you don't have to know what all the numbers are to begin with. You can use letters for the ones you don't know.

Sometimes, if there are enough clues, you can find out what numbers the letters stand for. It's a bit like a mystery. Say a pie is missing from your front stoop. It's been stolen by an unknown thief, whom we might call *X*. But there are clues. A trail of crumbs leads from your stoop to a neighbor's doghouse. The dog has crumbs on his whiskers and a self-satisfied look. The clues—the facts in the case—seem to fit the dog. So, *X* might = dog.

In algebra the clues are numbers. They are the facts we know—facts that can lead to the unknown.

For example, let's suppose your friend takes
another bag of jellybeans out of her pocket.
"Let's share these," she says.

But the only thing in the bag is a hole. The bag
tore while it was in her pocket—and that's where
the beans are now. So your friend fishes them out
and you count them. There are fifteen.

"We can't split all of those," she says. "I had seven extra beans in my pocket before the bag broke, that I promised to my sister. We can only split the ones that were in the bag."

How many were in the bag? Well, there are some clues. Let's put them into the language of algebra.

We can call the number of jellybeans in the bag before it broke *B*. And your friend had seven extra ones. That makes *B* + 7.

Altogether, she has fifteen. So $B + 7 = 15$. This is an EQUATION. The amount on one side of the "equals" sign is the same as the amount on the other side.

Is this true no matter what number *B* stands for? Is it true if *B* is 1? 2? 3? 4?

How can we find out what *B* is? We might try different numbers until we get one that makes the left side come out to 15.

$$B + 7 = 15$$
$$2 + 7 \neq 15$$
$$4 + 7 \neq 15$$

But is there a quicker way? Suppose we could get B all by itself on the left side of the equation. Then the number on the right side would tell us what B is.

How can we change the left side of the equation so that only B is on that side? Can we add or subtract something from the left side that will do the job?

One important thing about equations is that if we change one side by a certain amount, we have to change the other side by the same amount. Otherwise the two sides won't be equal anymore.

Have you found *B*? What's your fair share of the jellybeans?

In this problem, there is only one answer that fits the clues. Only if we put that number in place of *B* will the equation be true. That number is the SOLUTION to the equation.

My fair share
is 4.

Sometimes there is more than one correct solution. But if you have enough clues, you can always find a solution to an equation—even if sometimes the solution is pretty funny-looking.

For instance, try putting this puzzle into algebra.

You have twice as many jellybeans as your friend. Together, the two of you have 999. How many jellybeans do you have?

That's right!

About the Author

Malcolm E. Weiss is the author of many books and magazine articles
for children on science. He lives in Whitefield, Maine, with his wife
and daughters Margot and Rebecca.

When asked about how he came to write this book, Mr. Weiss
replied: "While I was weeding our vegetable garden one day, Margot
asked me to explain one of those 'think of a number' guessing
games to her. As I did, I realized that it would be a good way to
introduce algebra to young readers. Margot did too—until she saw
the finished manuscript. Then she said, 'I wish you had used
gumdrops.' "

About the Illustrator

The illustrator of numerous books for children, Judith Hoffman
Corwin was born in Watertown, New York, and earned her Bachelor
of Fine Arts degree at the Pratt Institute in New York City.

She now lives in New York City with her husband and their young
son, Oliver.